```
┌─────────────────────────┐
│                         │
│      This Book          │
│     Belongs To:         │
│                         │
│     ──────────────      │
│                         │
│     ──────────────      │
│                         │
└─────────────────────────┘
```

Write the number one

1 1 1

one one one one

2

Write the number two

2 2 2

_____

_____

two two two

_____

_____

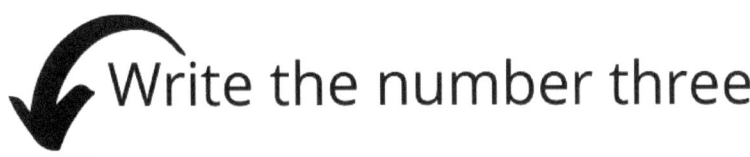 Write the number three

3 3 3

three three three
three

Write the number four

4 4 4

four four four

Write the number five

5 5 5

five five five
five

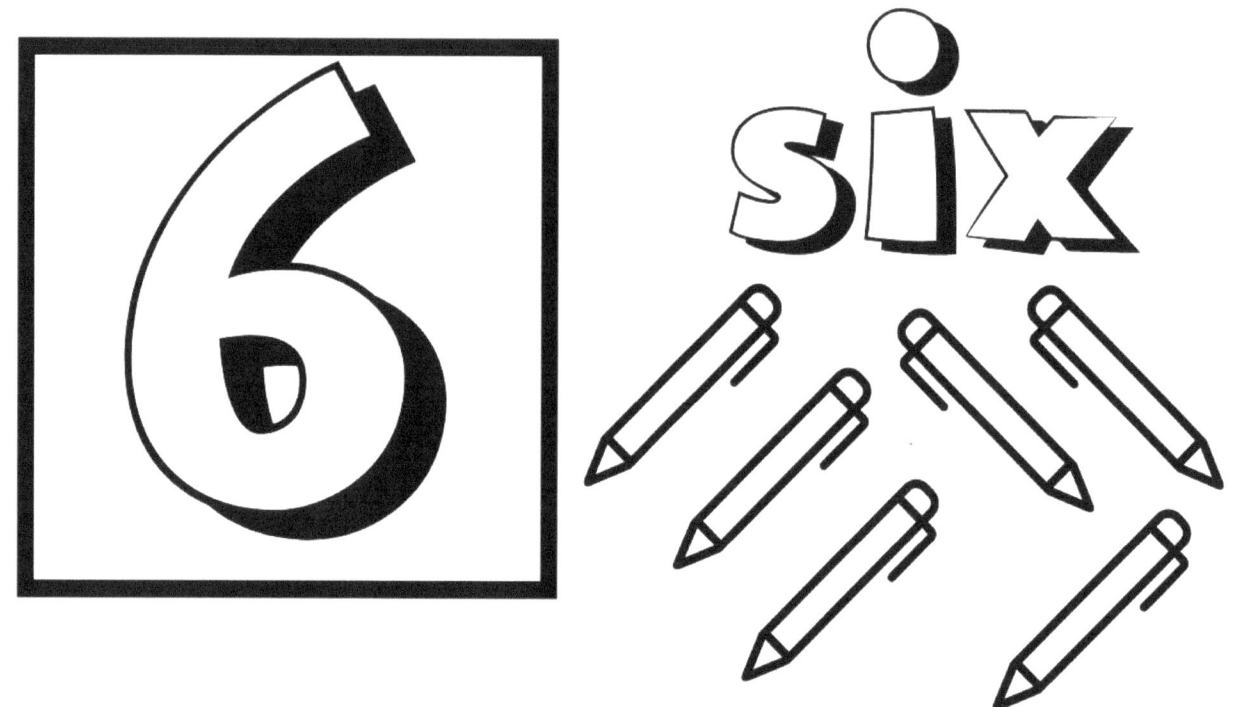

Write the number six

Write the number seven

7 7 7 7

seven seven
seven seven

Write the number eight

8 8 8

eight   eight
eight

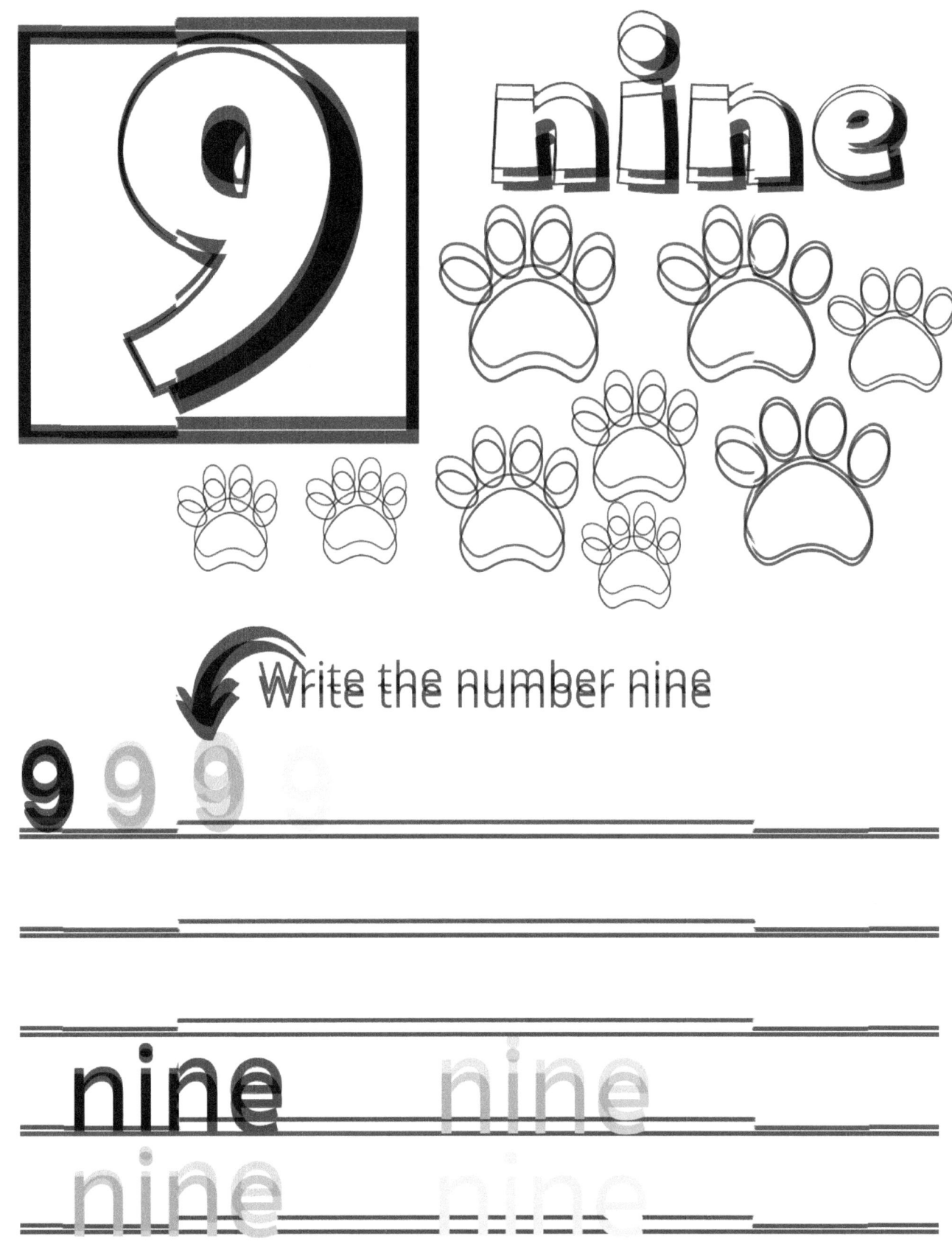

Write the number nine

9 9 9

nine   nine
nine   nine

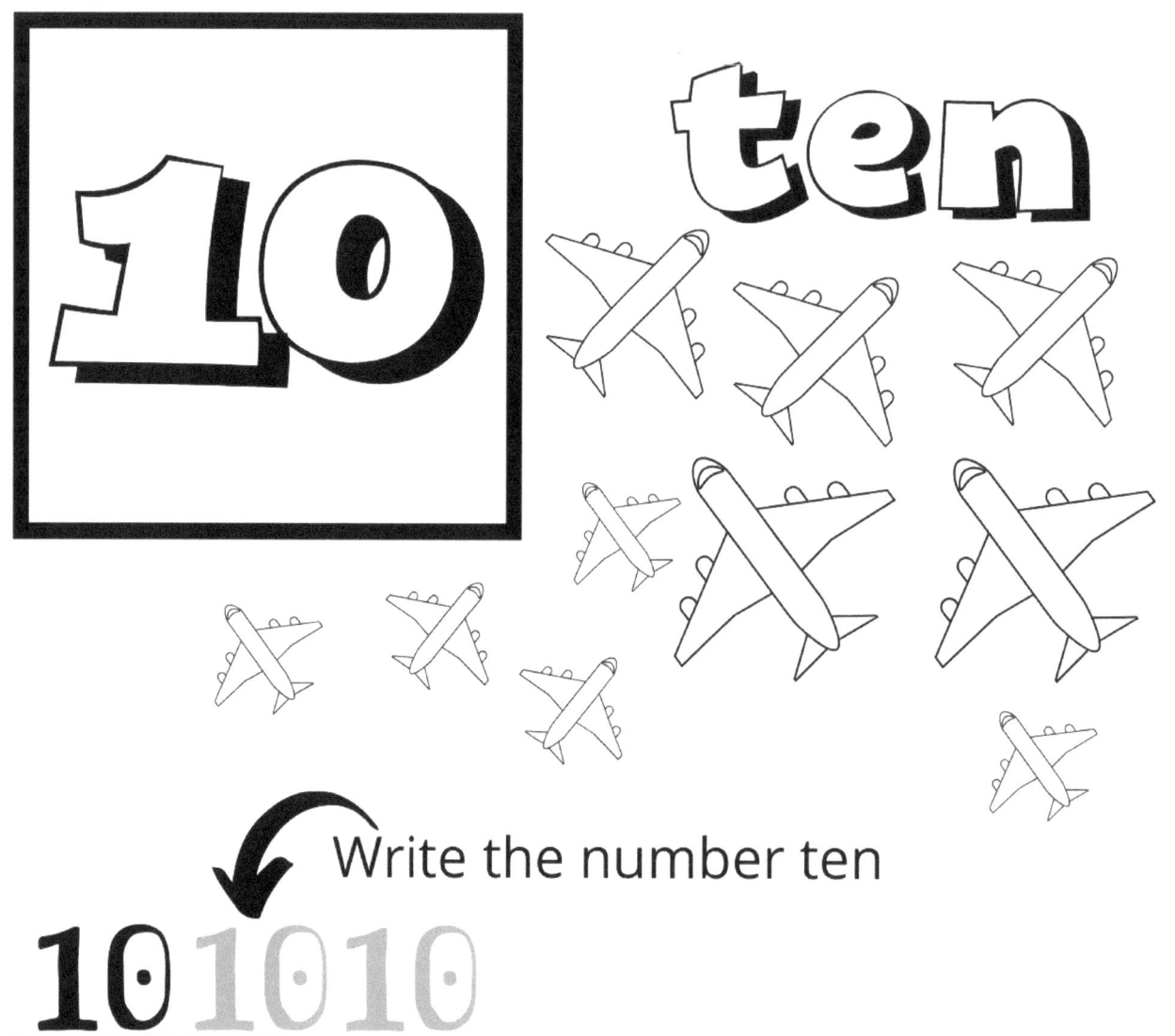

Write the number ten

10 10 10

ten ten ten

Missing Numbers

Can you fill in the missing numbers by launching rocket to the space?

Match the Number

Draw a line from the number to the matching set of objects:

2

4

3

5

8

Subtract Numbers

Find the spaceship. Subtract the numbers in each box and color the spaceship with the correct answer.

6 - 3 =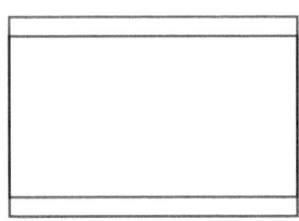

8 - 5 =

4 - 2 =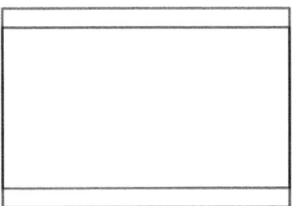

9 - 2 =

Coloring Addition

Find the answer tio each problem and write in the given box. Use the numbers to color the pictures.

6 + 3 =   Yellow

4 + 4 =   Green

2 + 2 =  Red

5 + 2 =  Blue

8 + 2 =  Orange

3 + 2 =  Pink

Counting

Count the objects. Write the correct number in the box. Color the objects for fun.

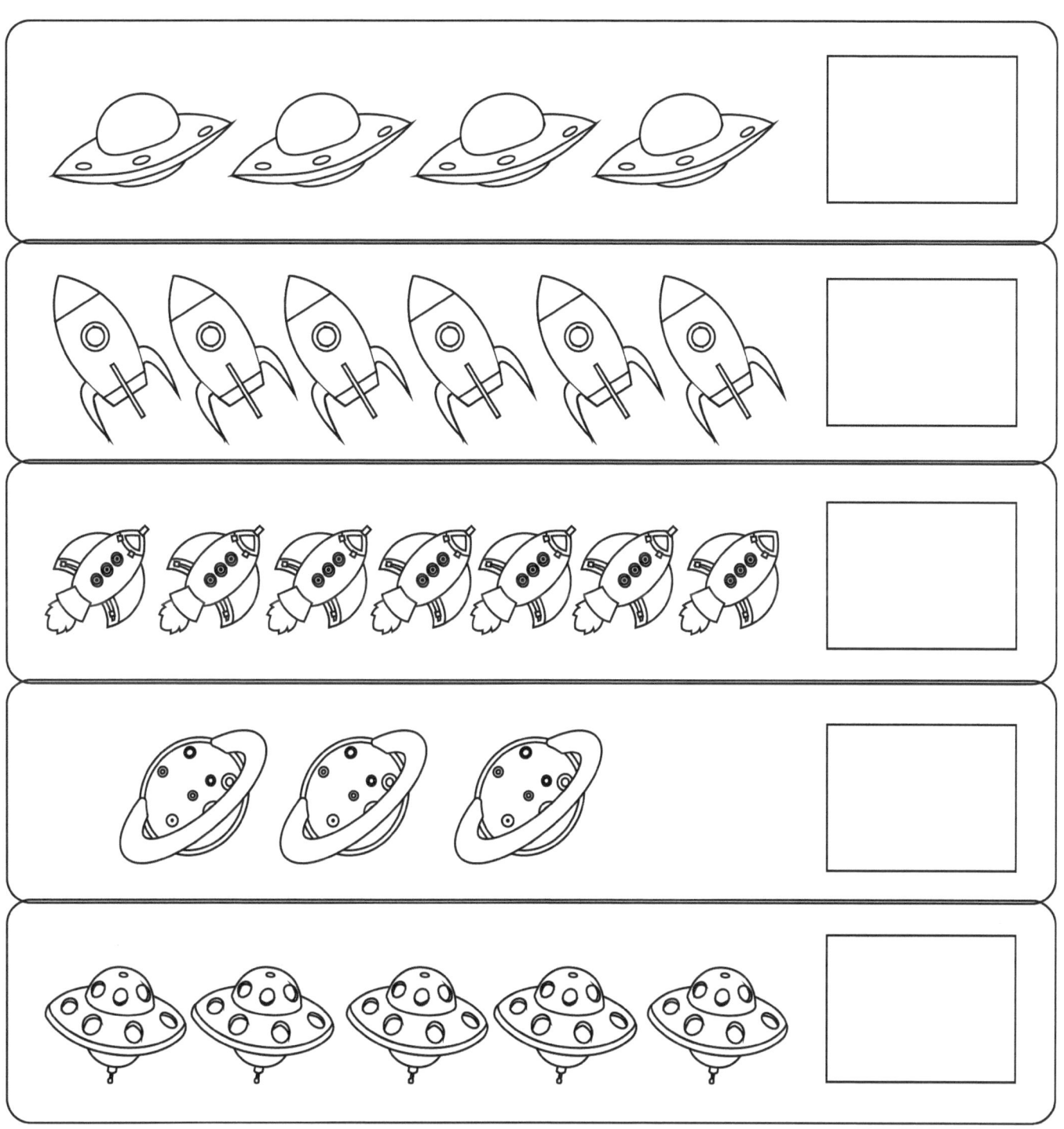

Connects the number to the image

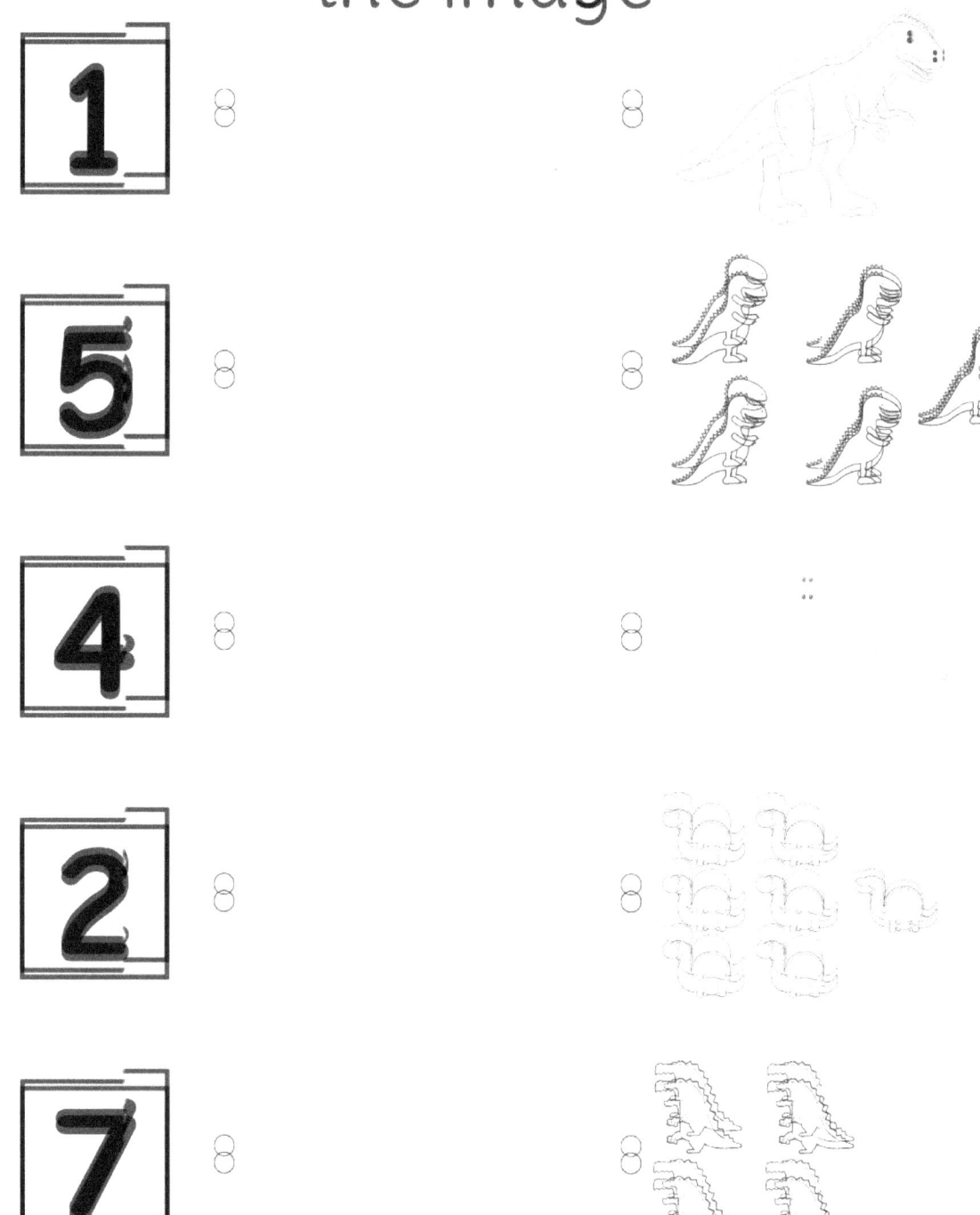

1 ○　　○

5 ○　　○

4 ○　　○

2 ○　　○

7 ○　　○

Connects the number to the image

1 ○   ○ 🦕🦕🦕

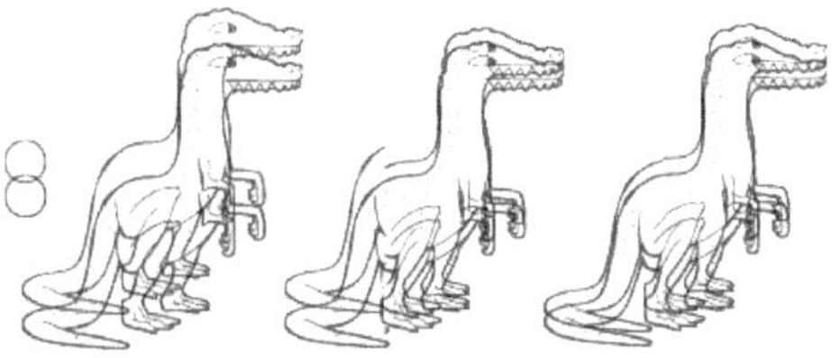

2 ○   ○ 🦕

3 ○   ○ ❀❀❀❀❀

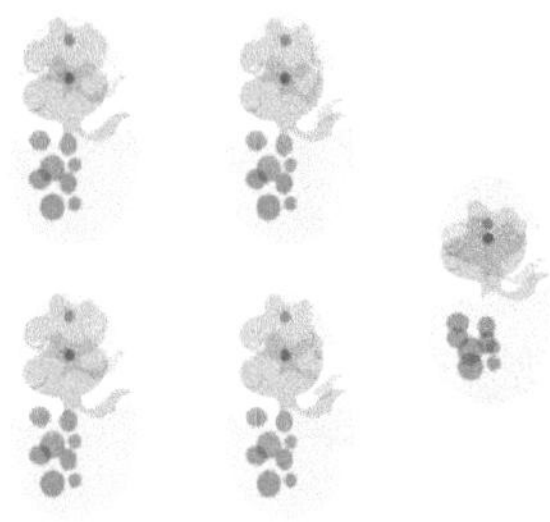

4 ○   ○ 🦖🦖🦖🦖

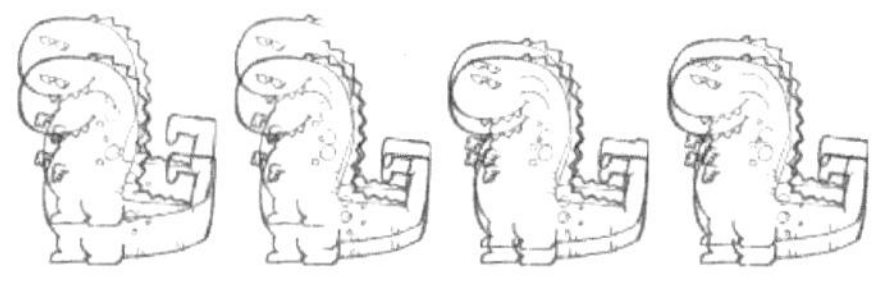

5 ○   ○ 🦏🦏

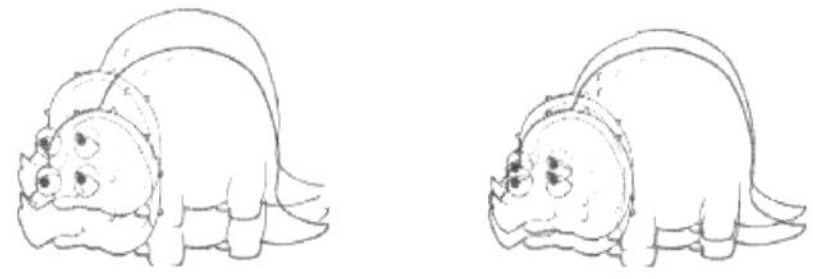

Connects the number to the image

2

8

5

7

1

Connects the number to the image

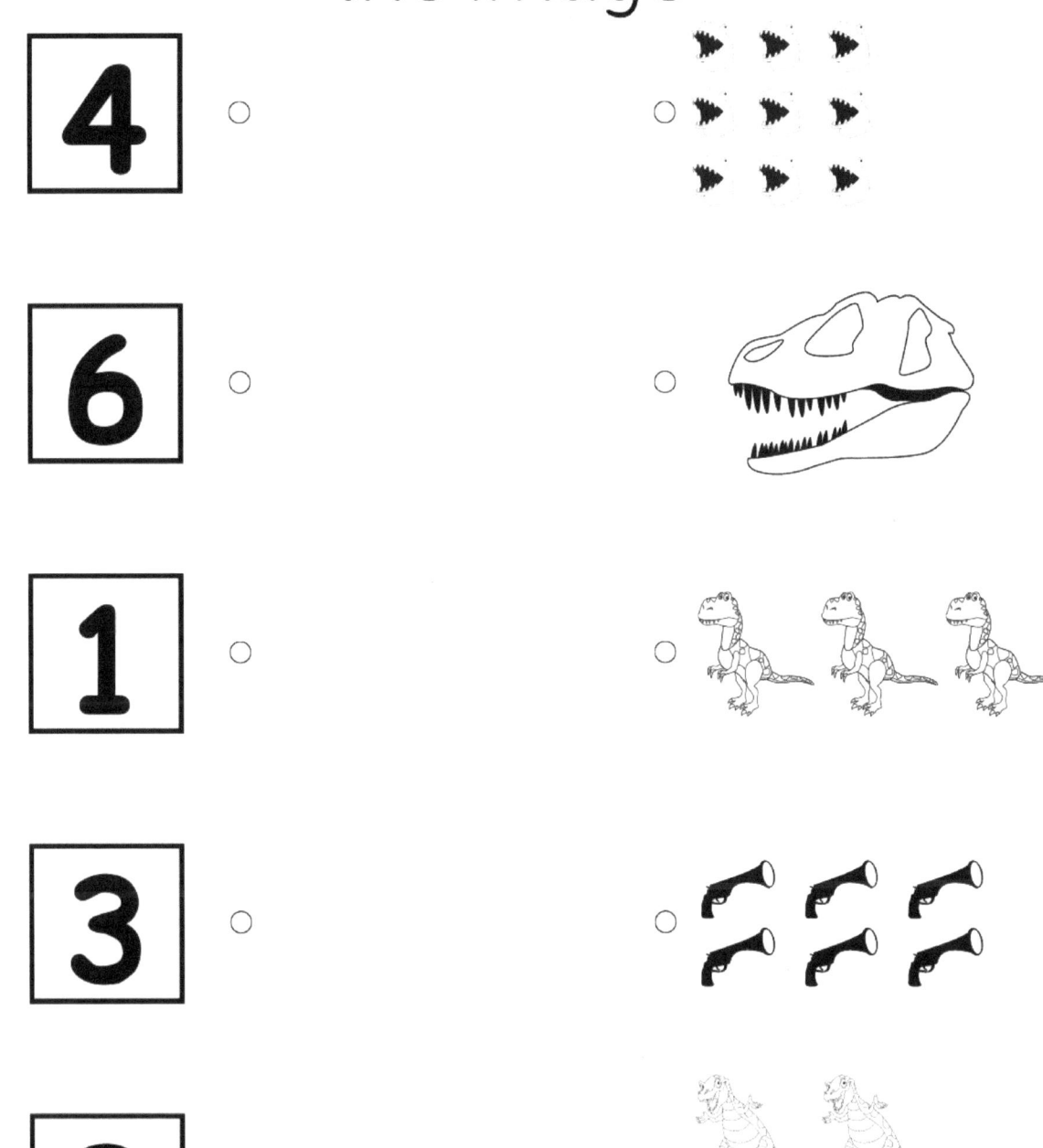

Connects the number to the image

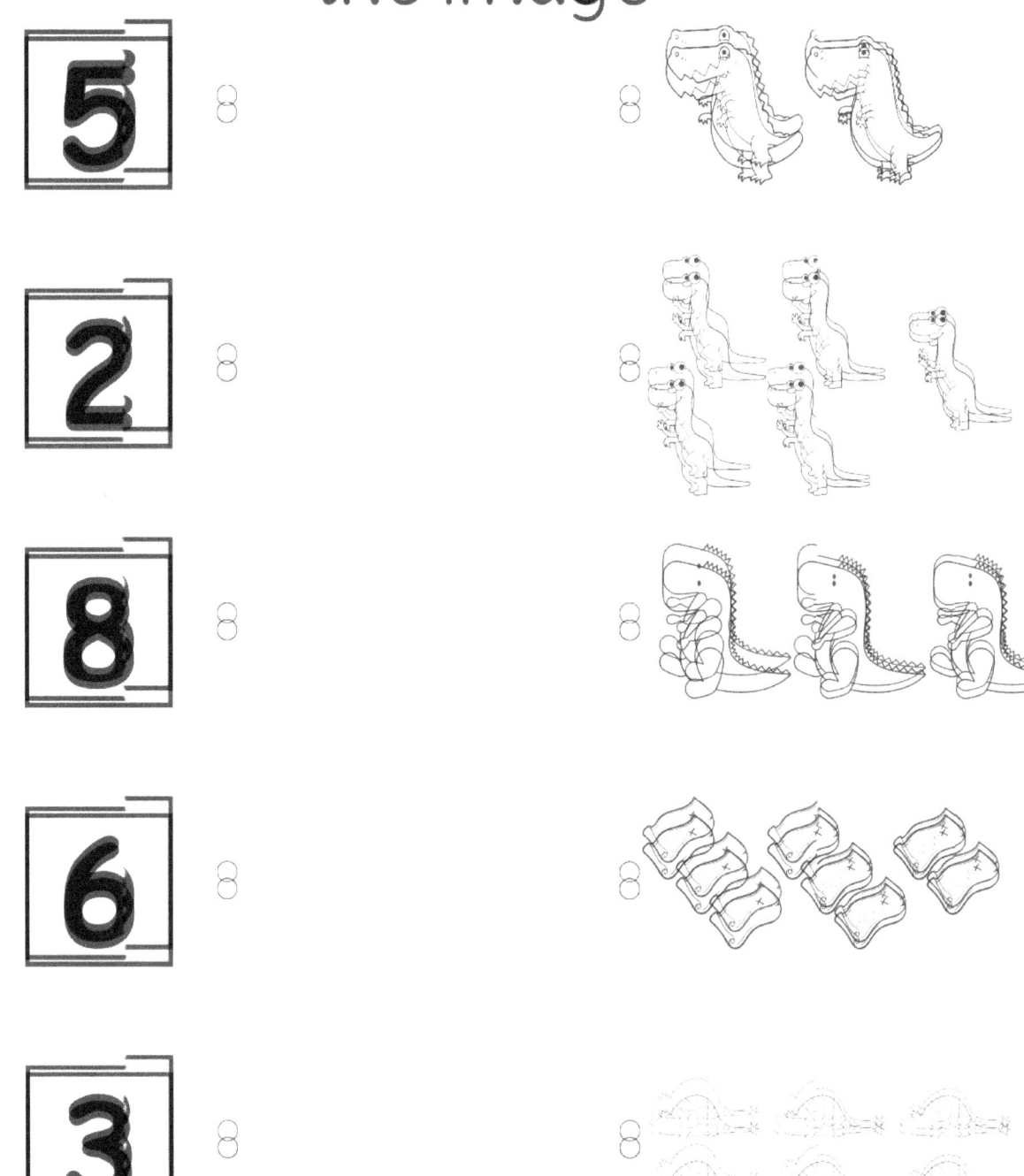

Magic Coloring

6 + 3 ≡ ☐ yellow   1 + 2 ≡ ☐ blue

2 + 5 ≡ ☐ green   1 + 4 ≡ ☐ orange

3 + 5 ≡ ☐ red   2 + 2 ≡ ☐ pink

Magic Coloring

6 + 3 = ☐ yellow   1 + 2 = ☐ blue

2 + 5 = ☐ green   1 + 4 = ☐ orange

3 + 5 = ☐ red   2 + 2 = ☐ pink

Magic Coloring

6 + 3 = ☐ yellow   1 + 2 = ☐ blue

2 + 5 = ☐ green   1 + 4 = ☐ orange

3 + 5 = ☐ red   2 + 2 = ☐ pink

Magic Coloring

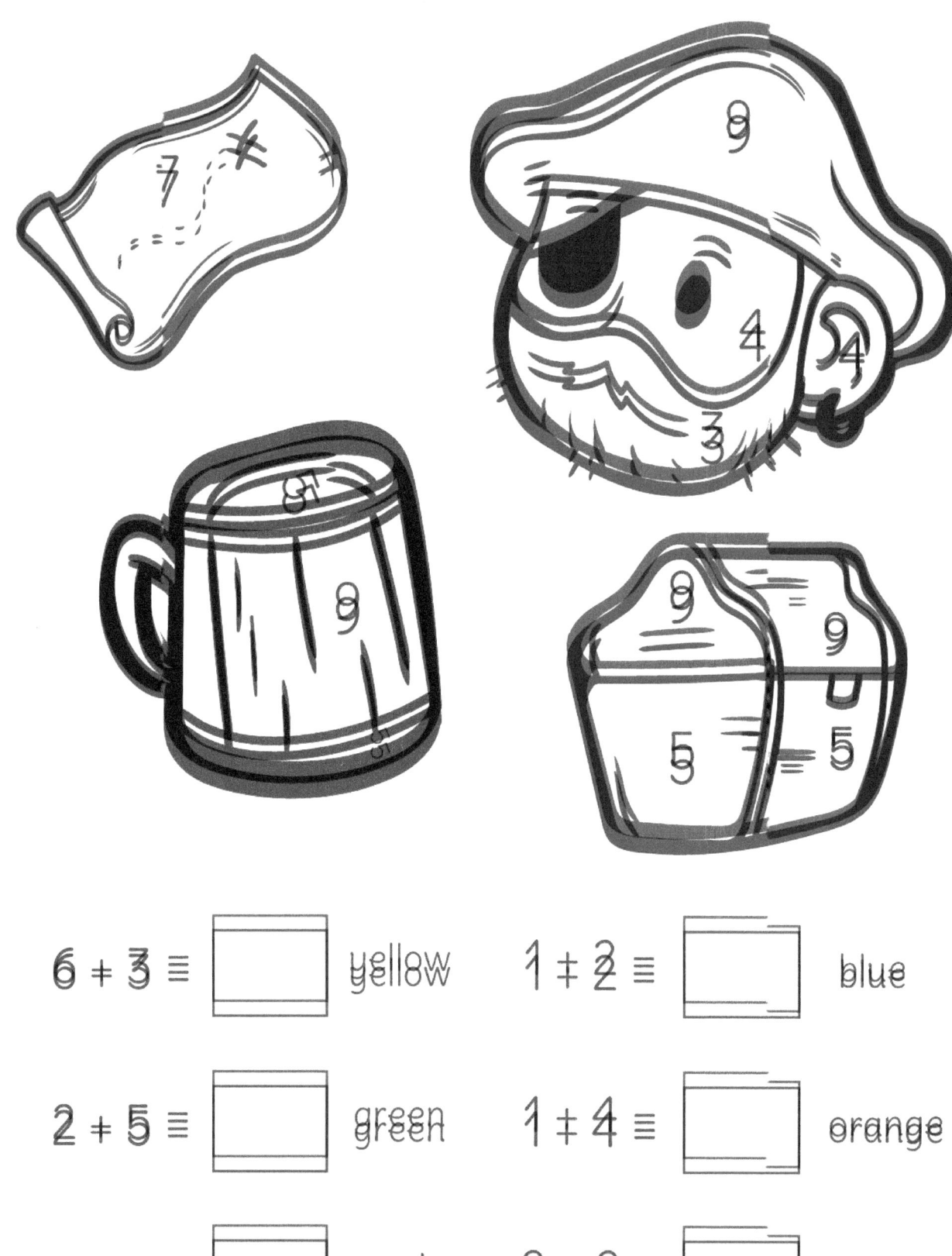

$6 + 3 =$ ☐ yellow $1 + 2 =$ ☐ blue

$2 + 5 =$ ☐ green $1 + 4 =$ ☐ orange

$3 + 5 =$ ☐ red $2 + 2 =$ ☐ pink

Magic Coloring

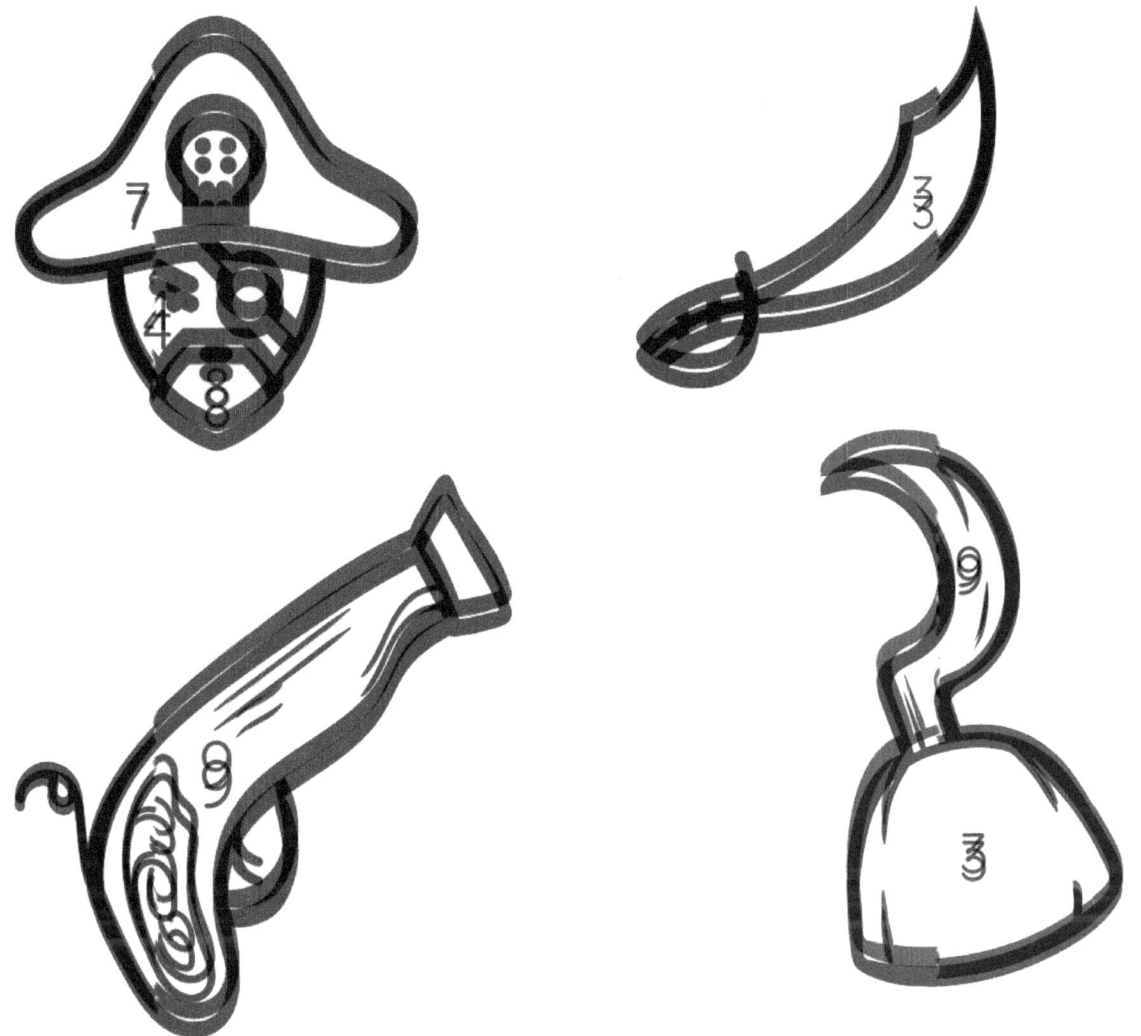

6 + 3 = ☐   yellow   1 + 2 = ☐   blue

2 + 5 = ☐   green   1 + 4 = ☐   orange

3 + 5 = ☐   red   2 + 2 = ☐   pink

MATH WORKSHEET
ADDING

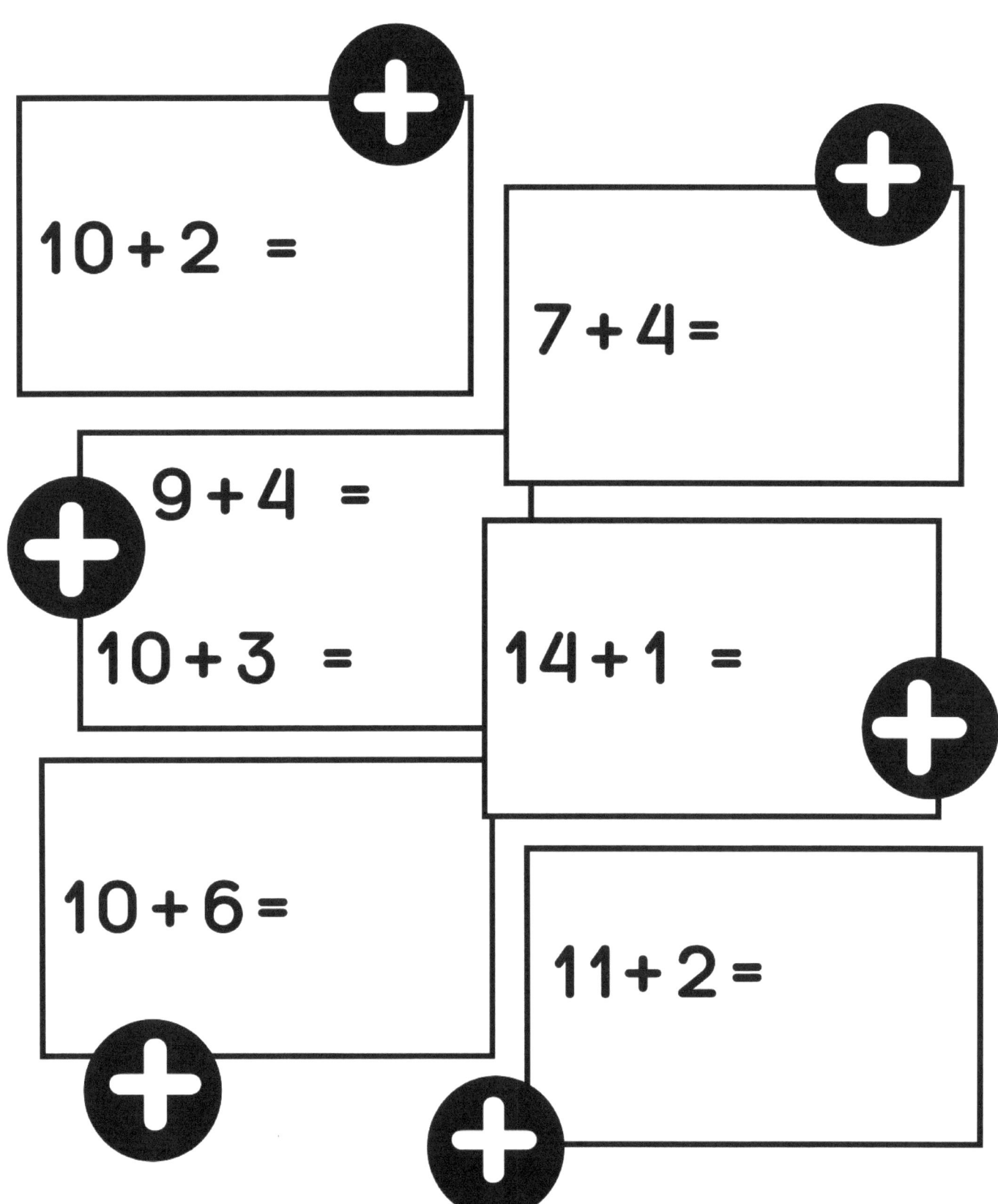

MATH WORKSHEET
SUBTRACTING

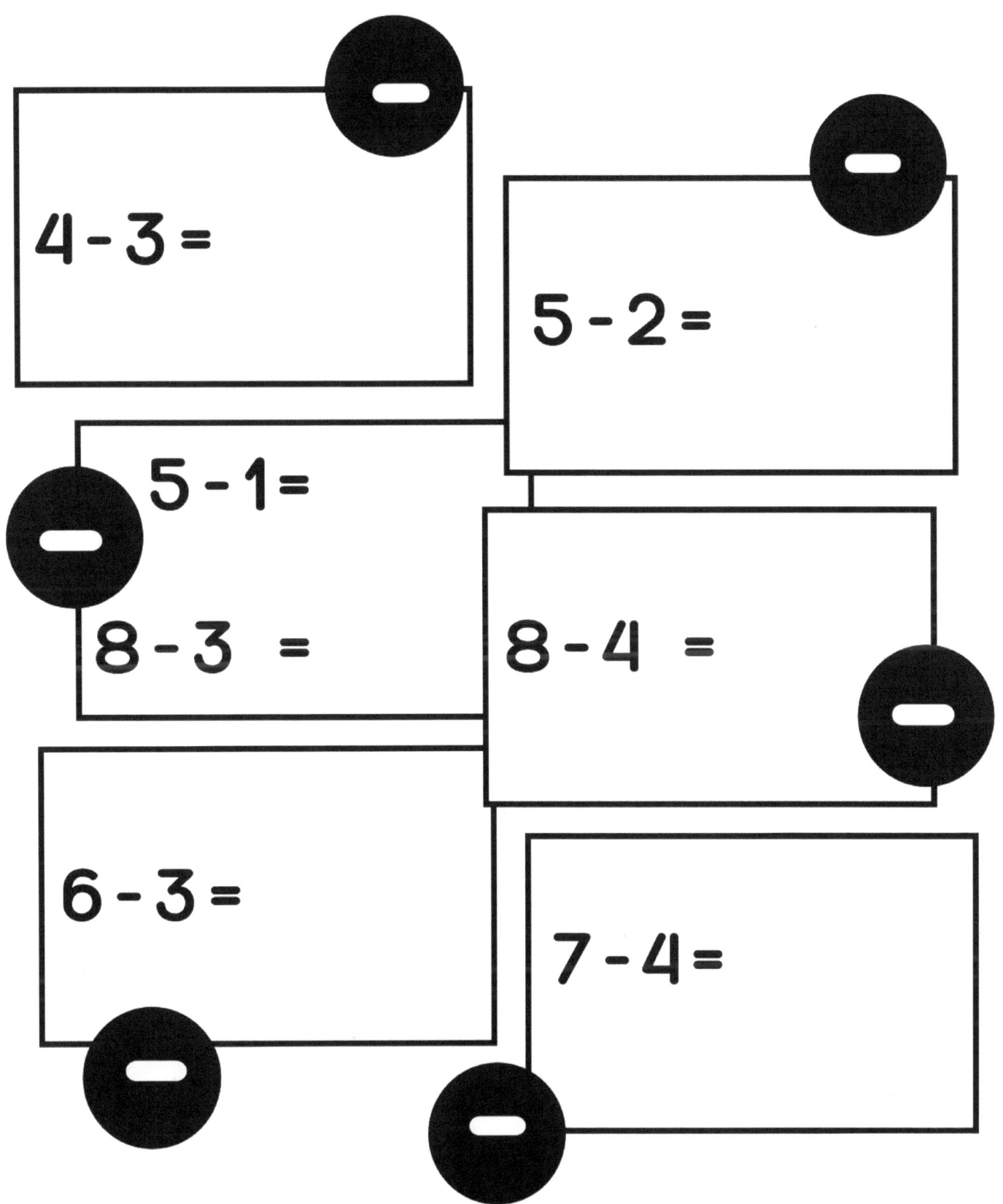

MATH WORKSHEET
ADDING

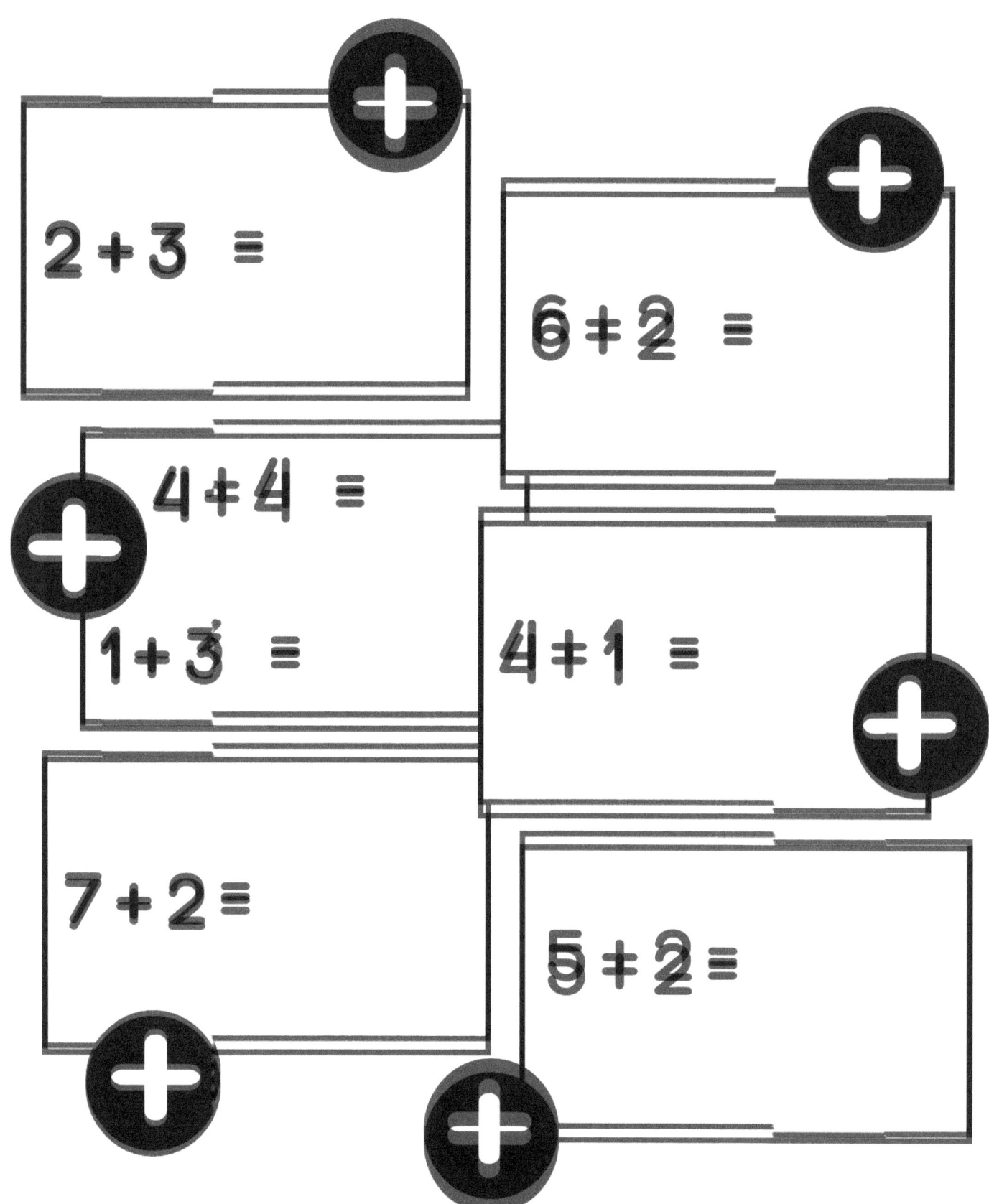

2 + 3 =

6 + 2 =

4 + 4 =

1 + 3 =

4 + 1 =

7 + 2 =

5 + 2 =

MATH WORKSHEET
SUBTRACTING

25 - 5 =

22 - 6 =

20 - 5 =

20 - 10 =

19 - 8 =

16 - 12 =

18 - 11 =

MATH WORKSHEET
ADDING

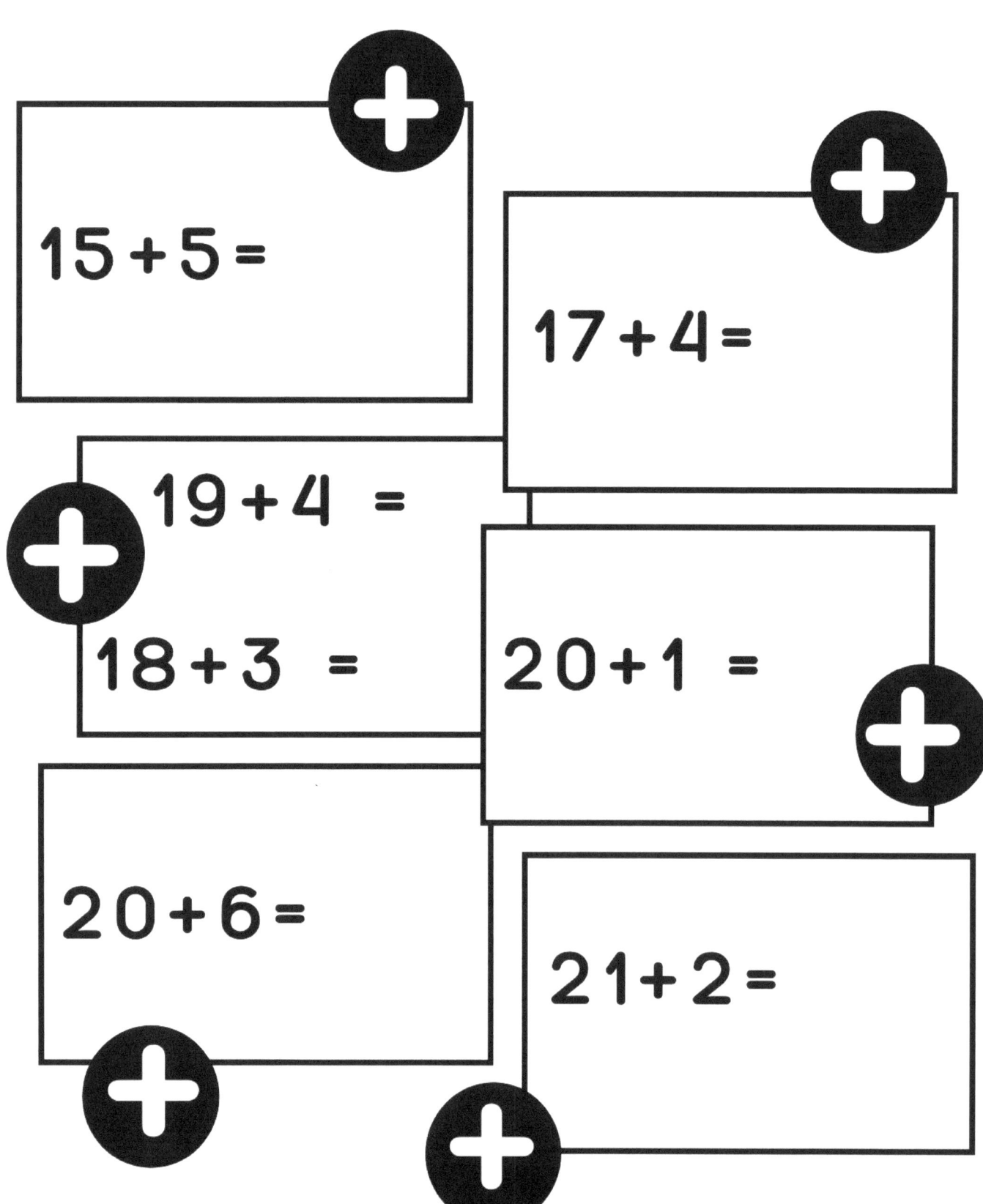

15 + 5 =

17 + 4 =

19 + 4 =

18 + 3 =

20 + 1 =

20 + 6 =

21 + 2 =

MATH WORKSHEET
SUBTRACTING

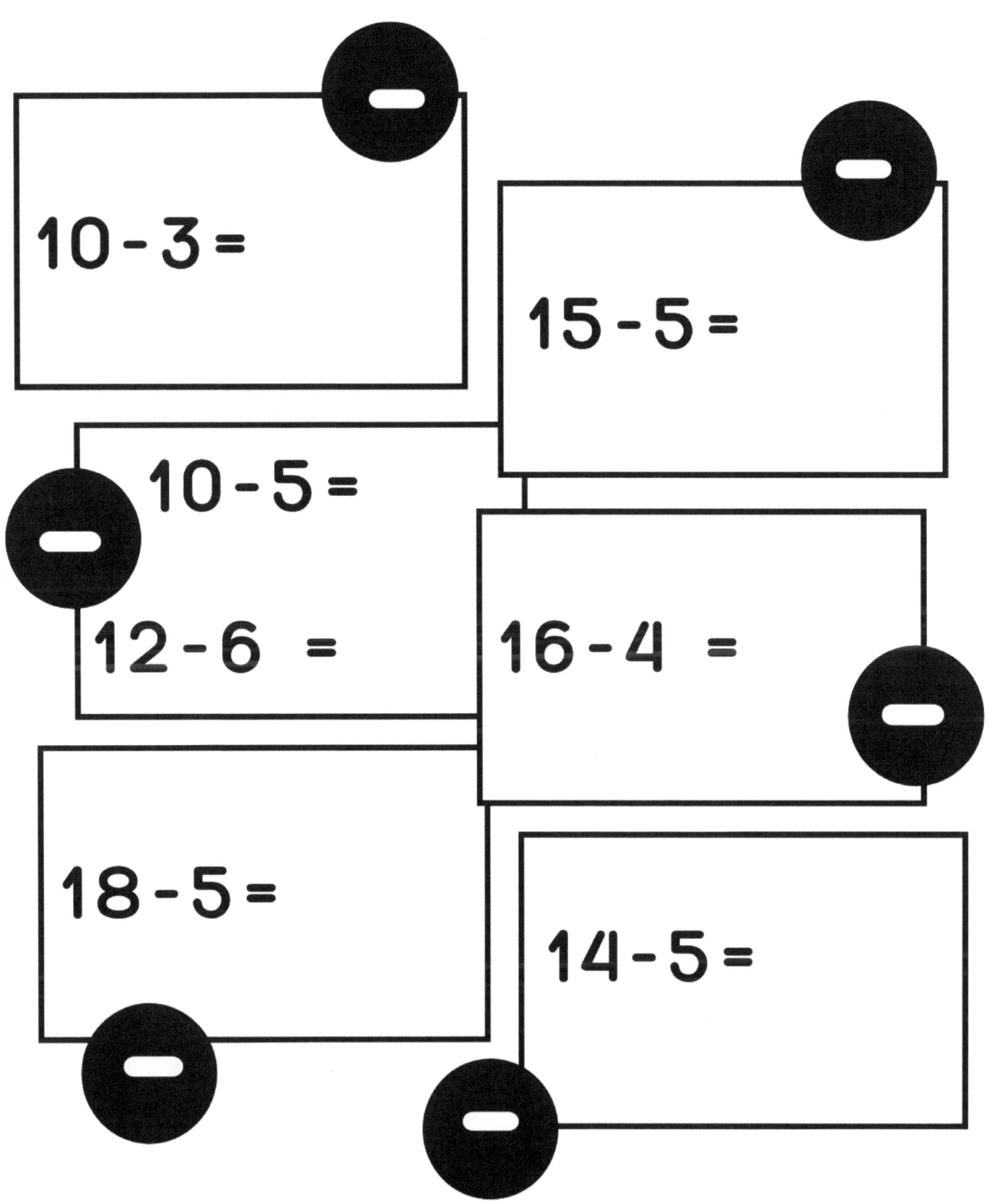

MATH WORKSHEET
TRACE, COUNT, WRITE

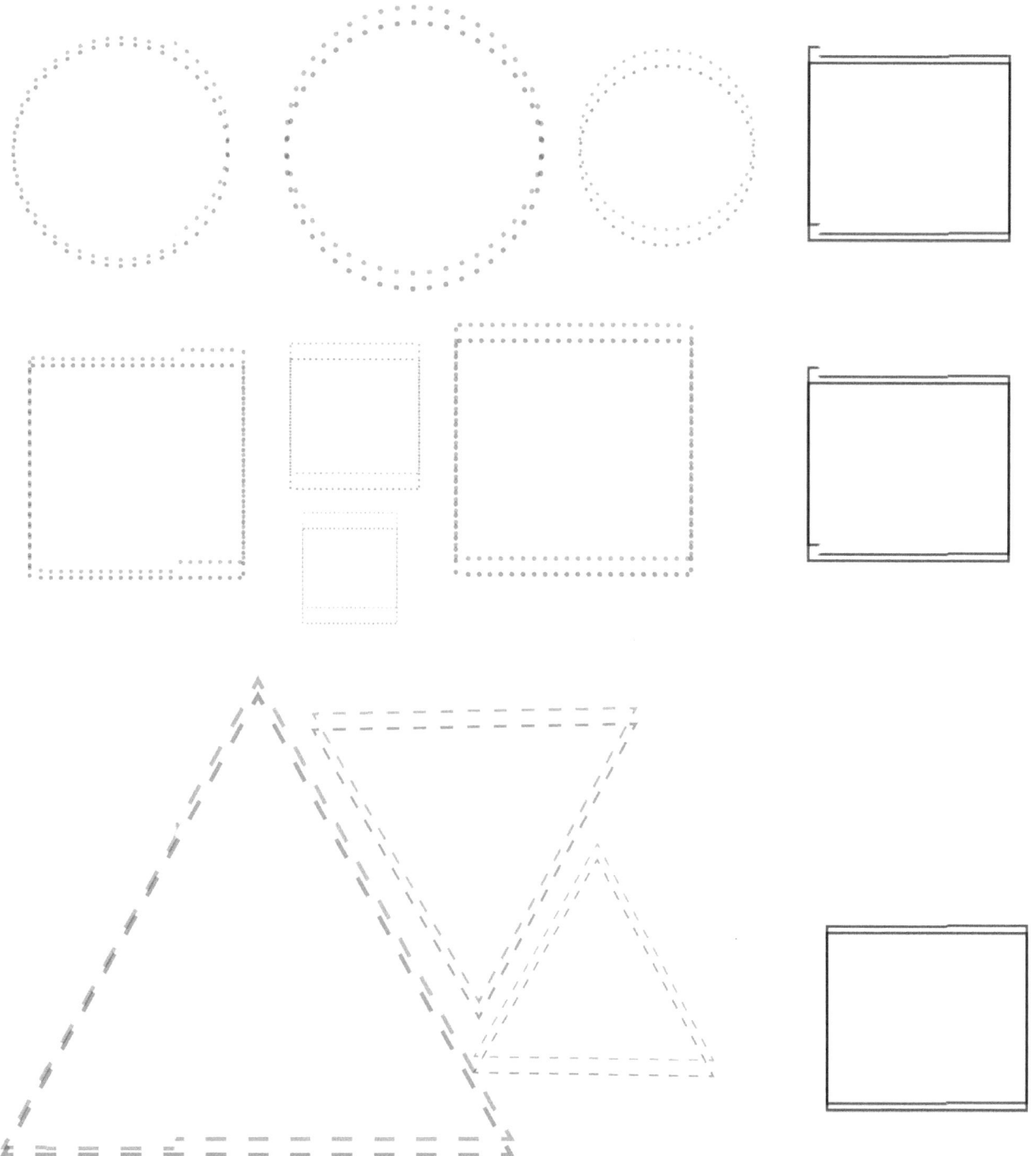

MATH WORKSHEET
WRITE THE MISSING NUMBER

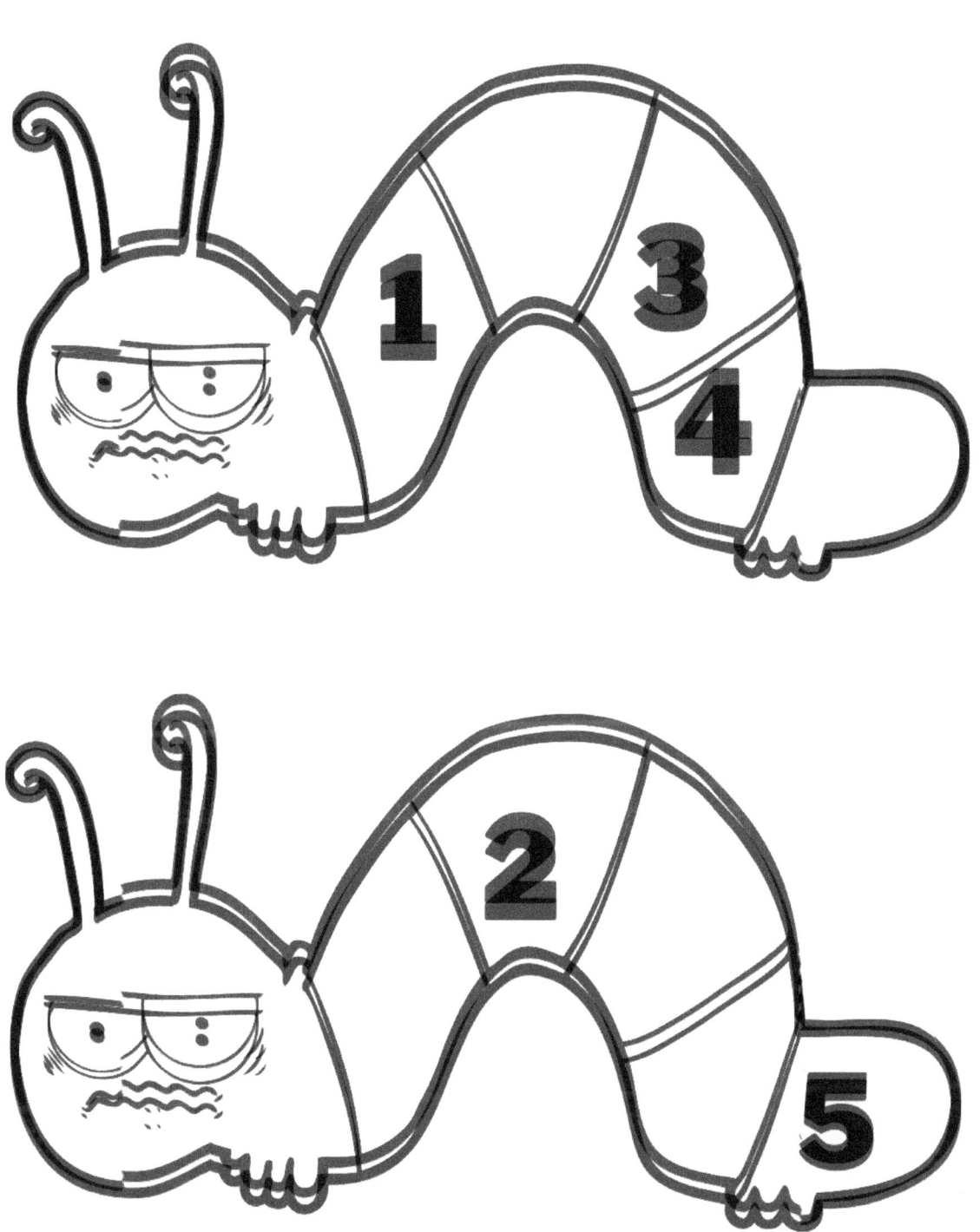

MATH WORKSHEET
COUNT THE OBJECTS AND WRITE THE NUMBER

Easter Egg
Coloring Activity

SHAPES

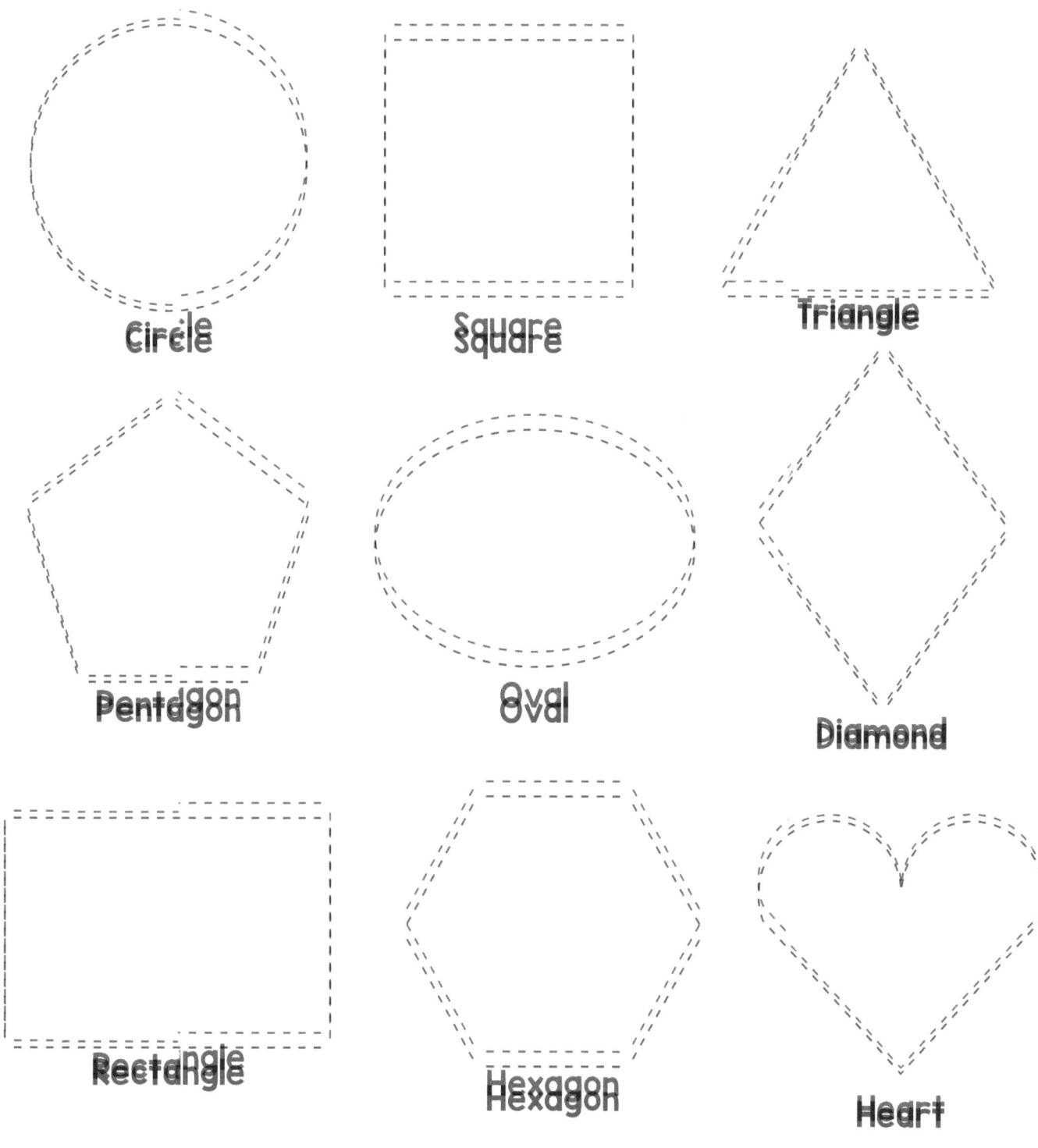

CIRCLE THE PICTURE THAT IS DIFFERENT

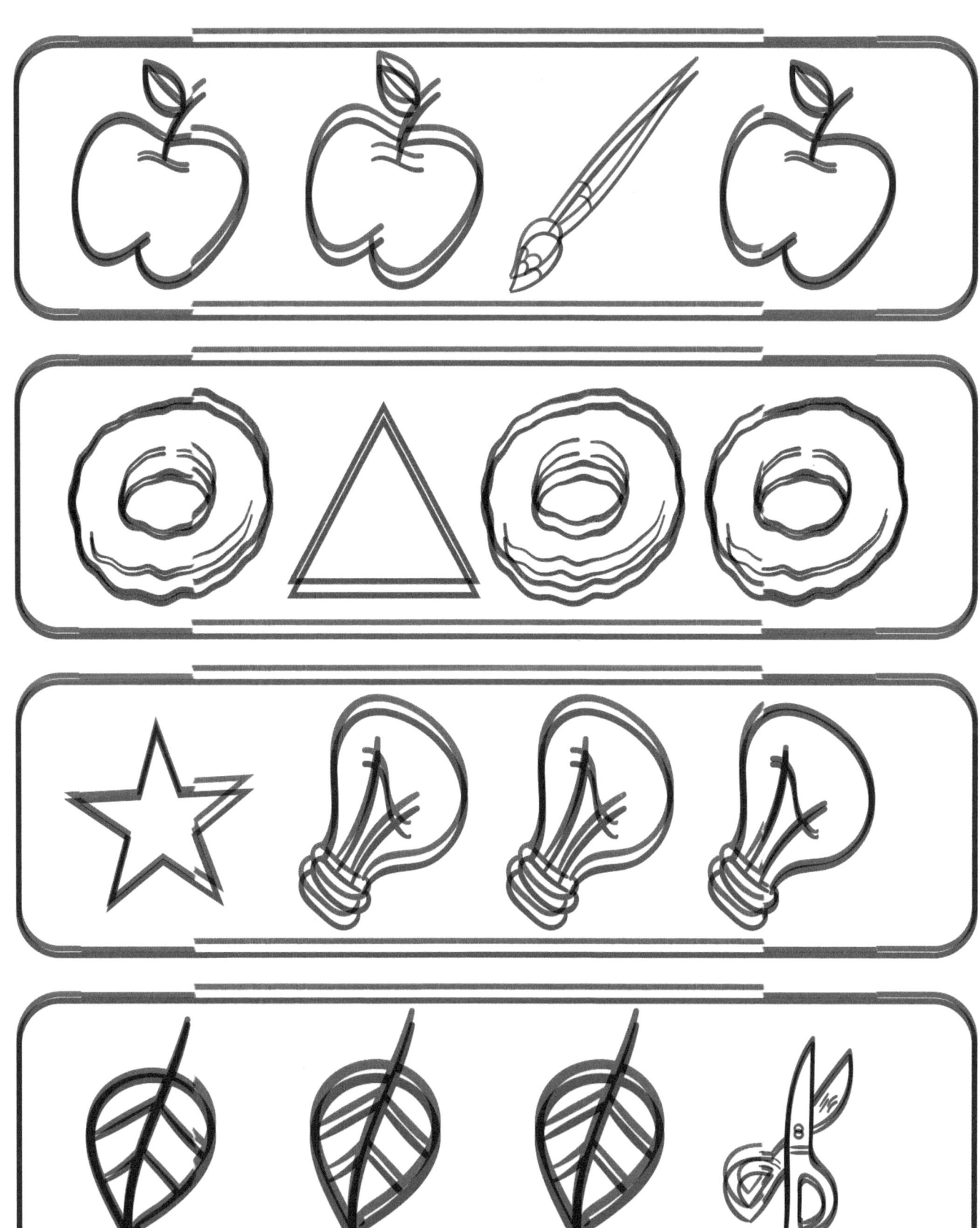

HOW MANY ARE THERE?
Connect the number with the same amount of objects

CIRCLE THE NUMBER
Circle the number that matches the amount of objects

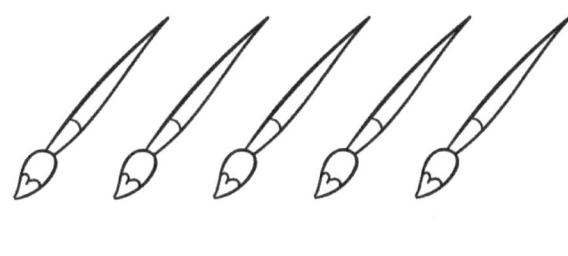

4   5   6   7

3   6   9   12

4   5   6   7

2   4   6   8

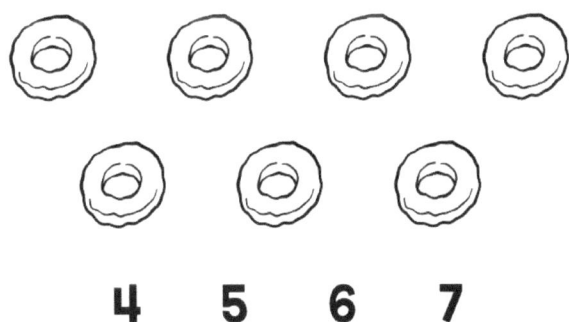

4   5   6   7

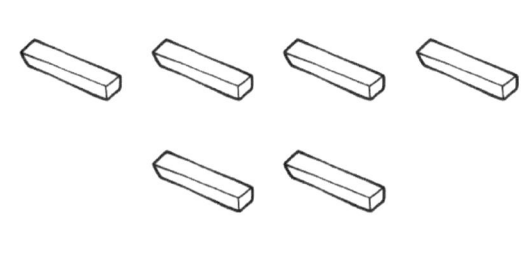

2   4   6   8

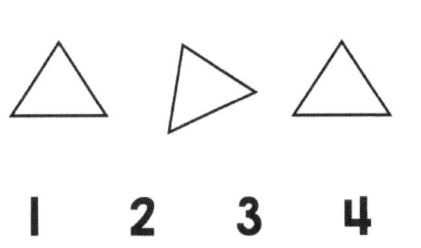

1   2   3   4

1   2   3   4

FIND THE NUMBER 1

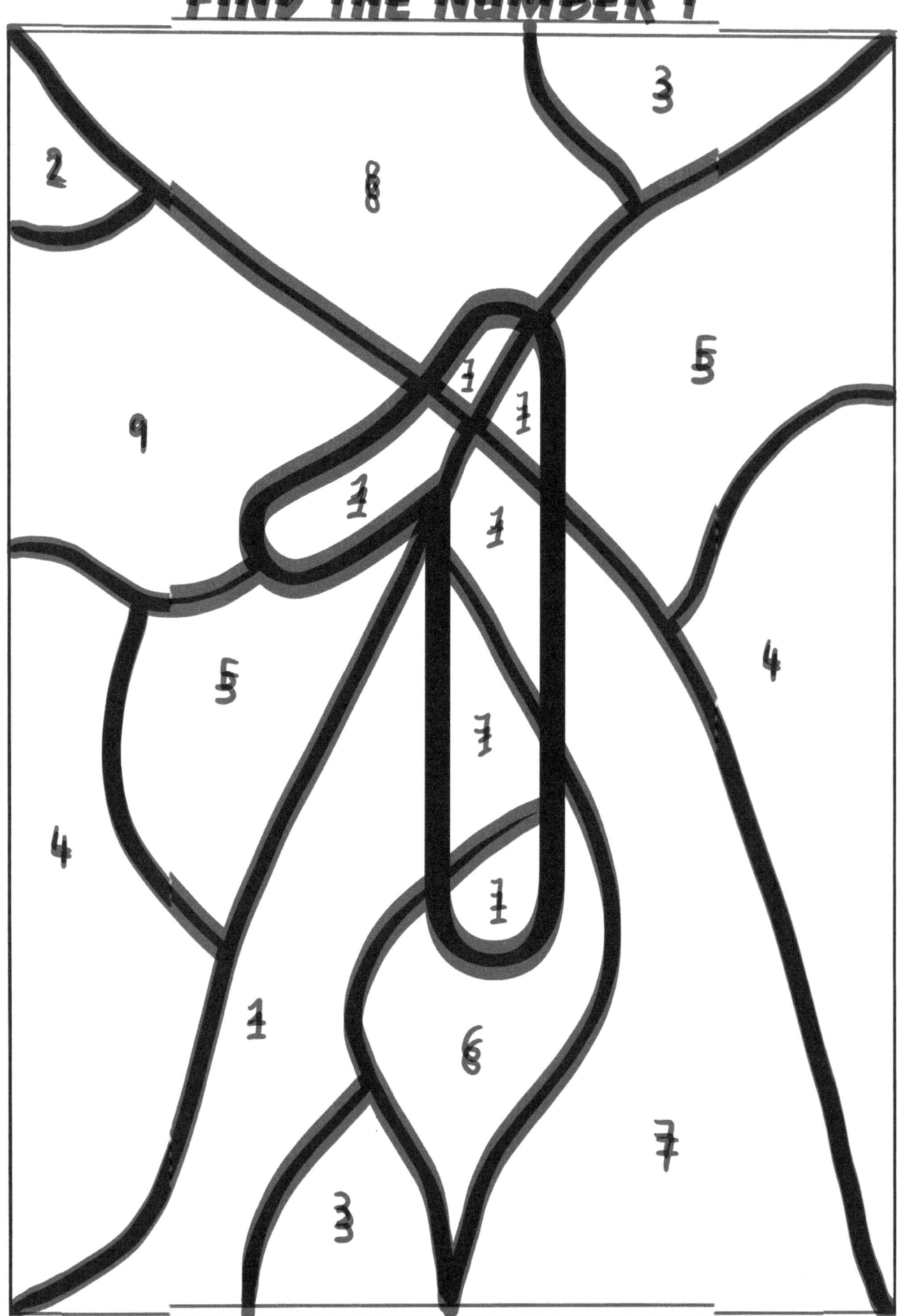

FIND THE NUMBER 2

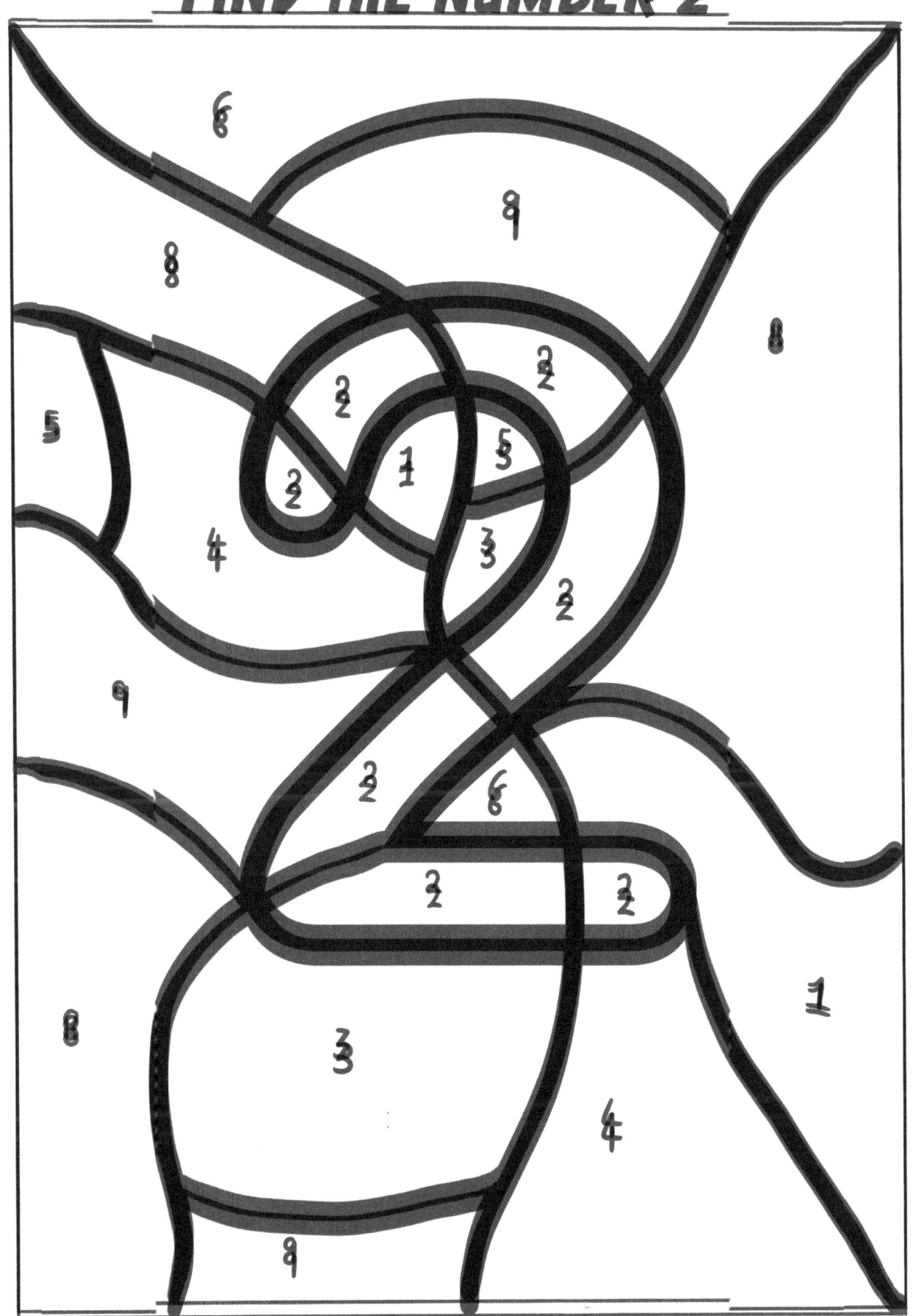

FIND THE NUMBER 3

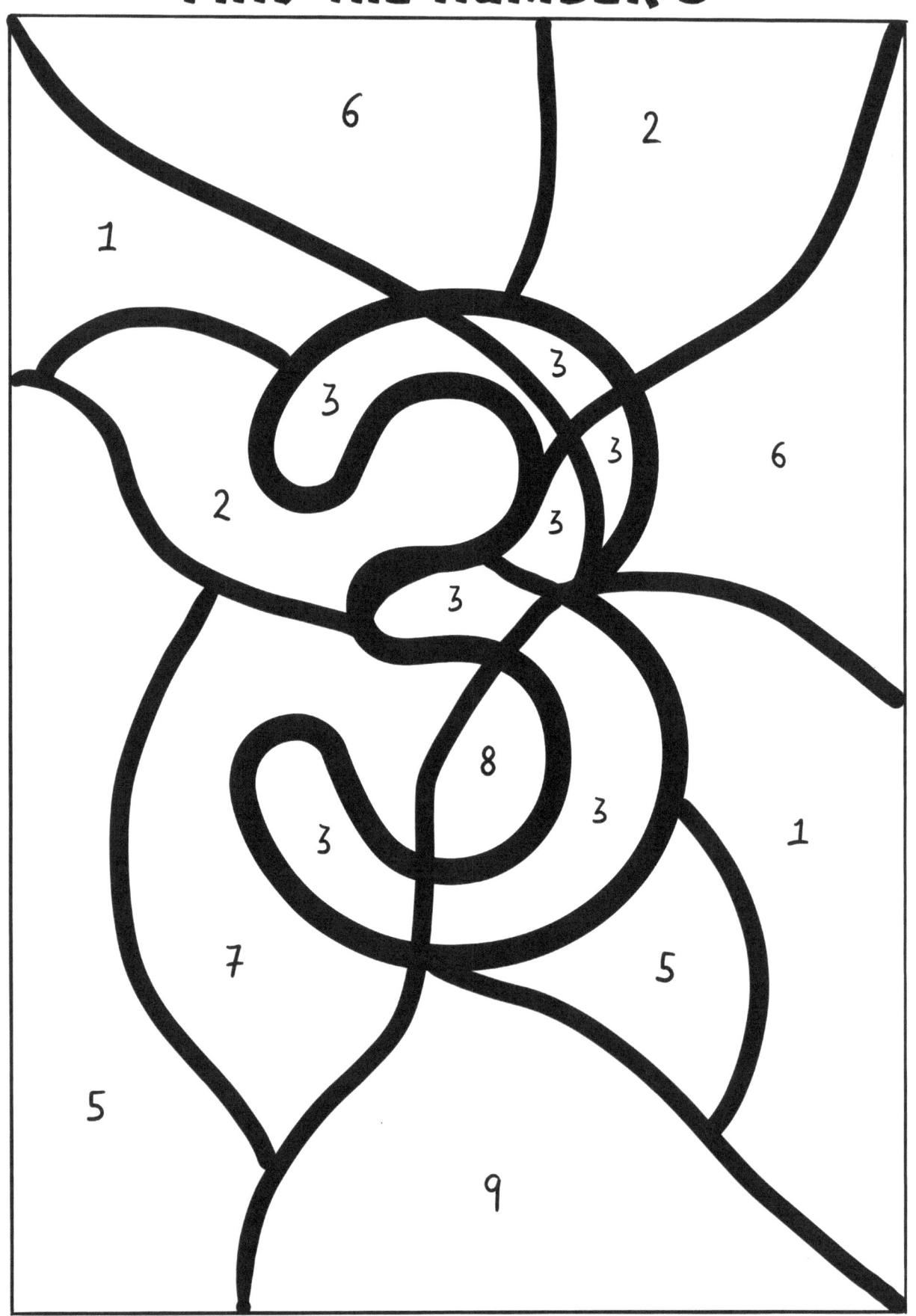

FIND THE NUMBER 4

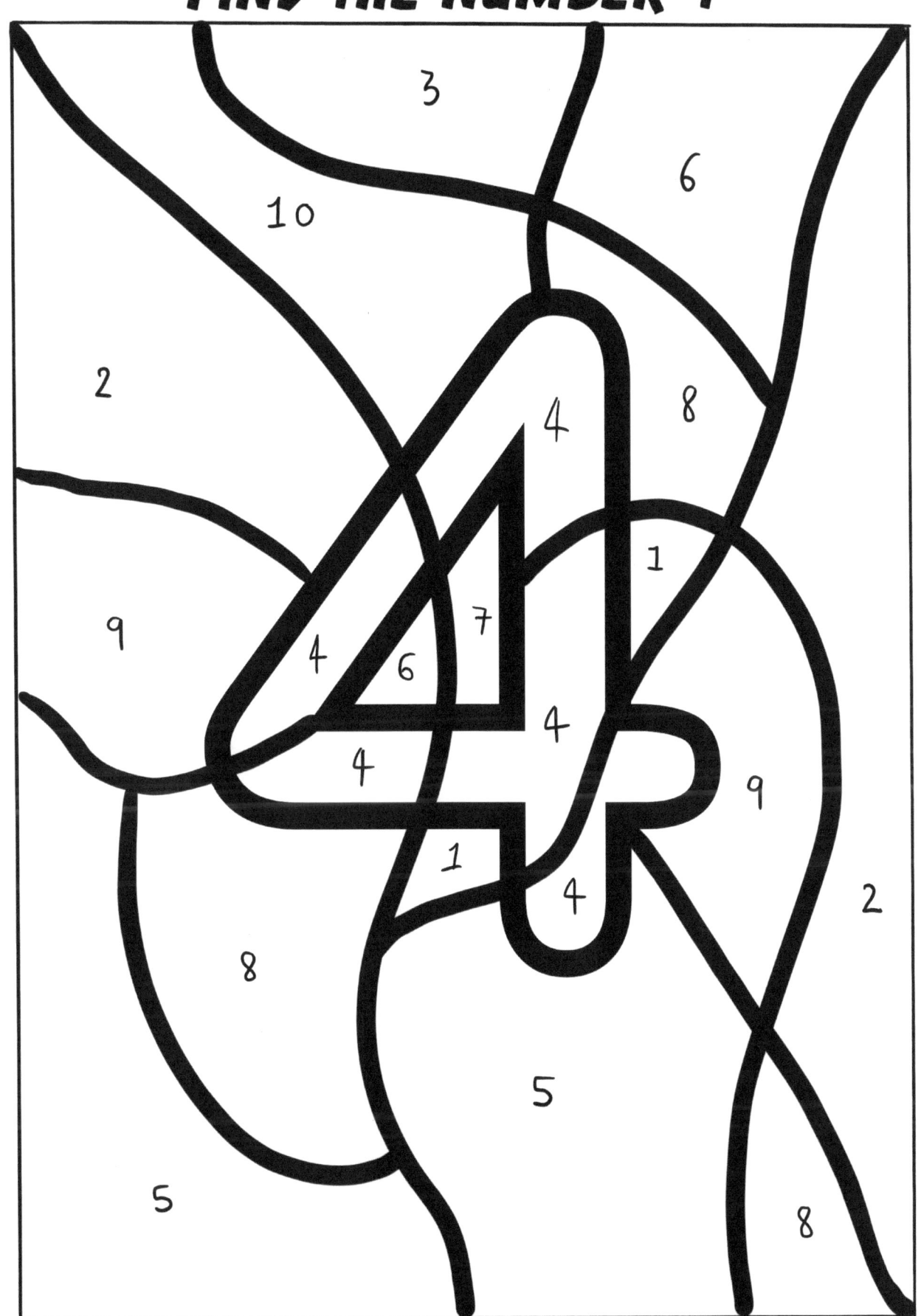

FIND THE NUMBER 5

FIND THE NUMBER 6

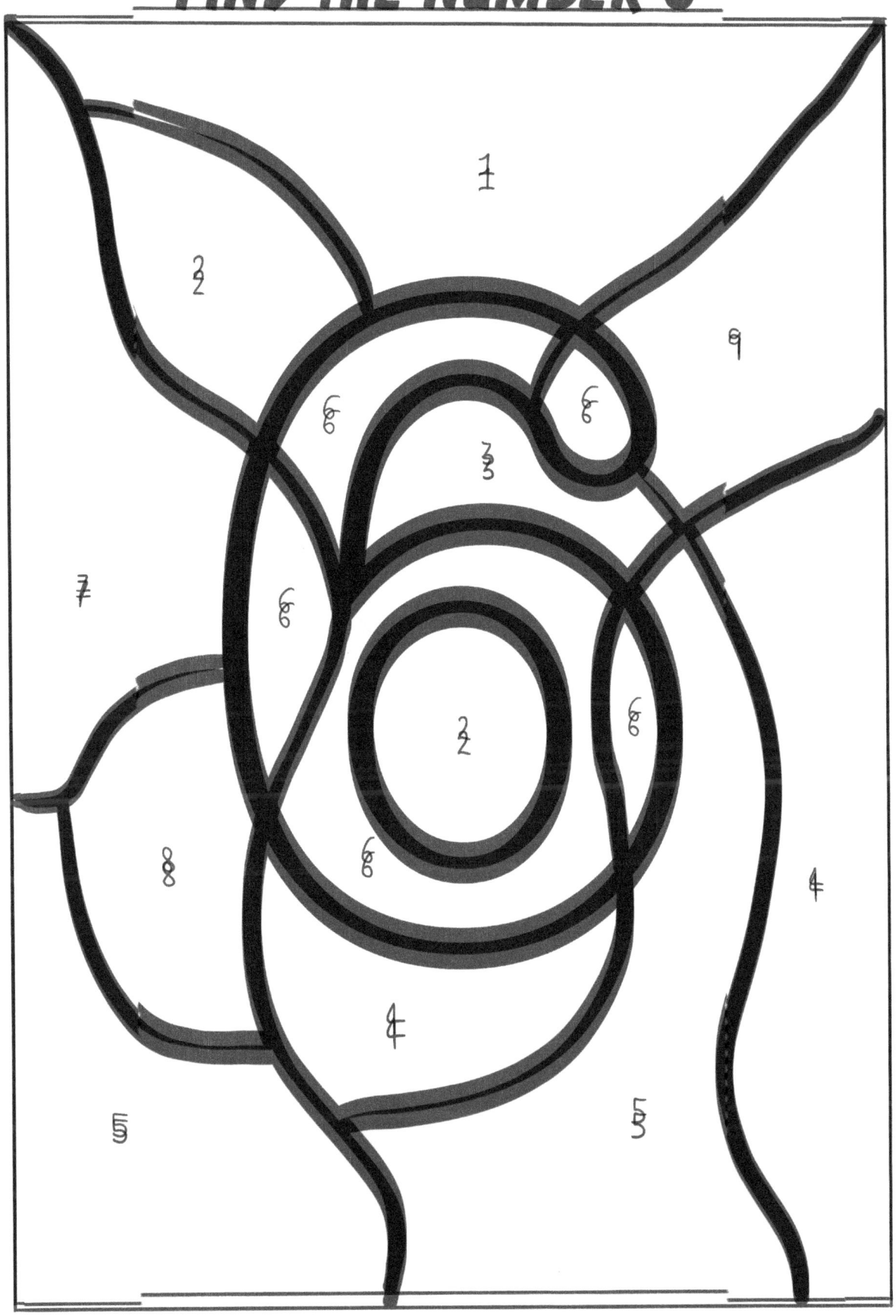

FIND THE NUMBER 7

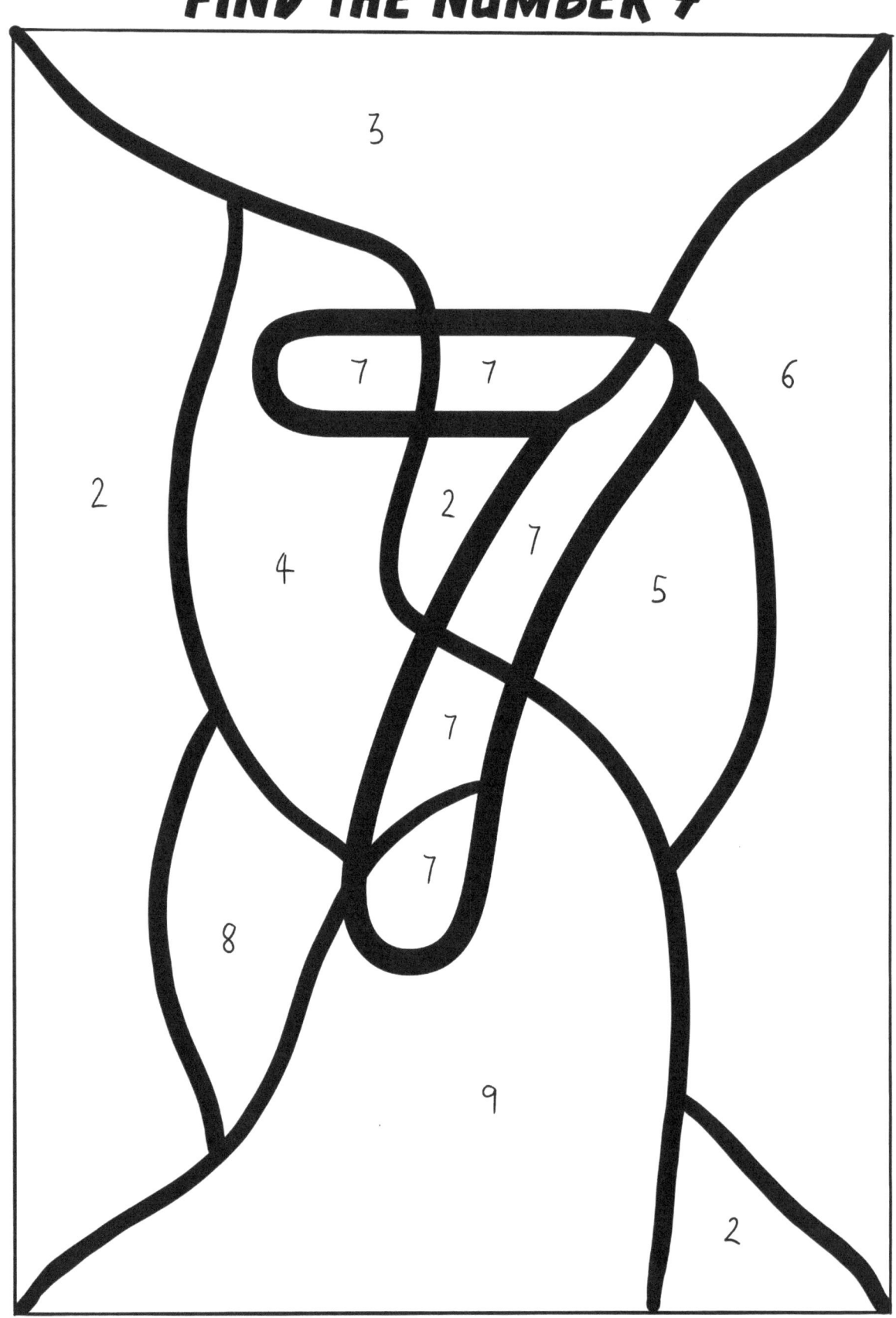

FIND THE NUMBER 8

FIND THE NUMBER 9

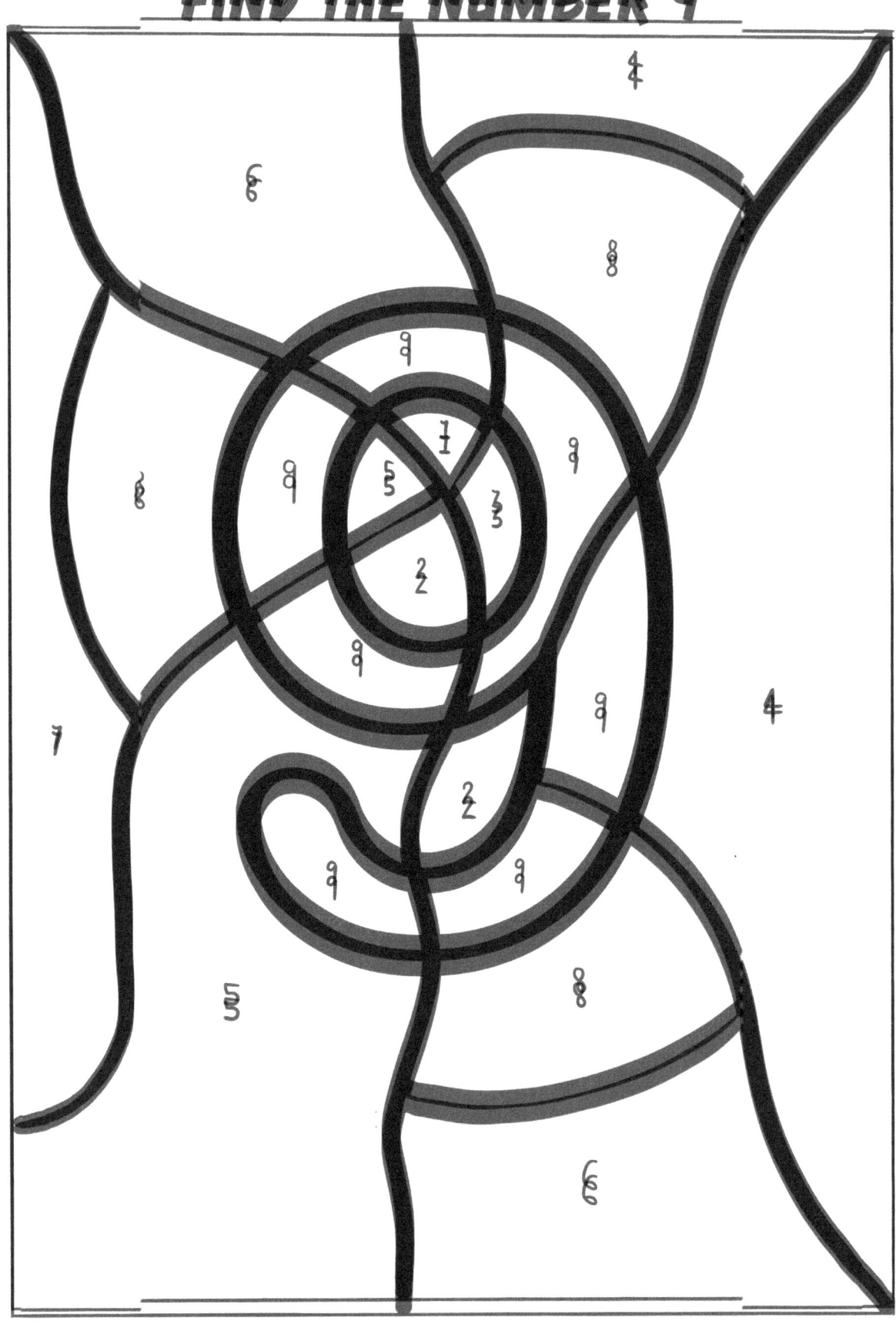

FIND THE NUMBER 10

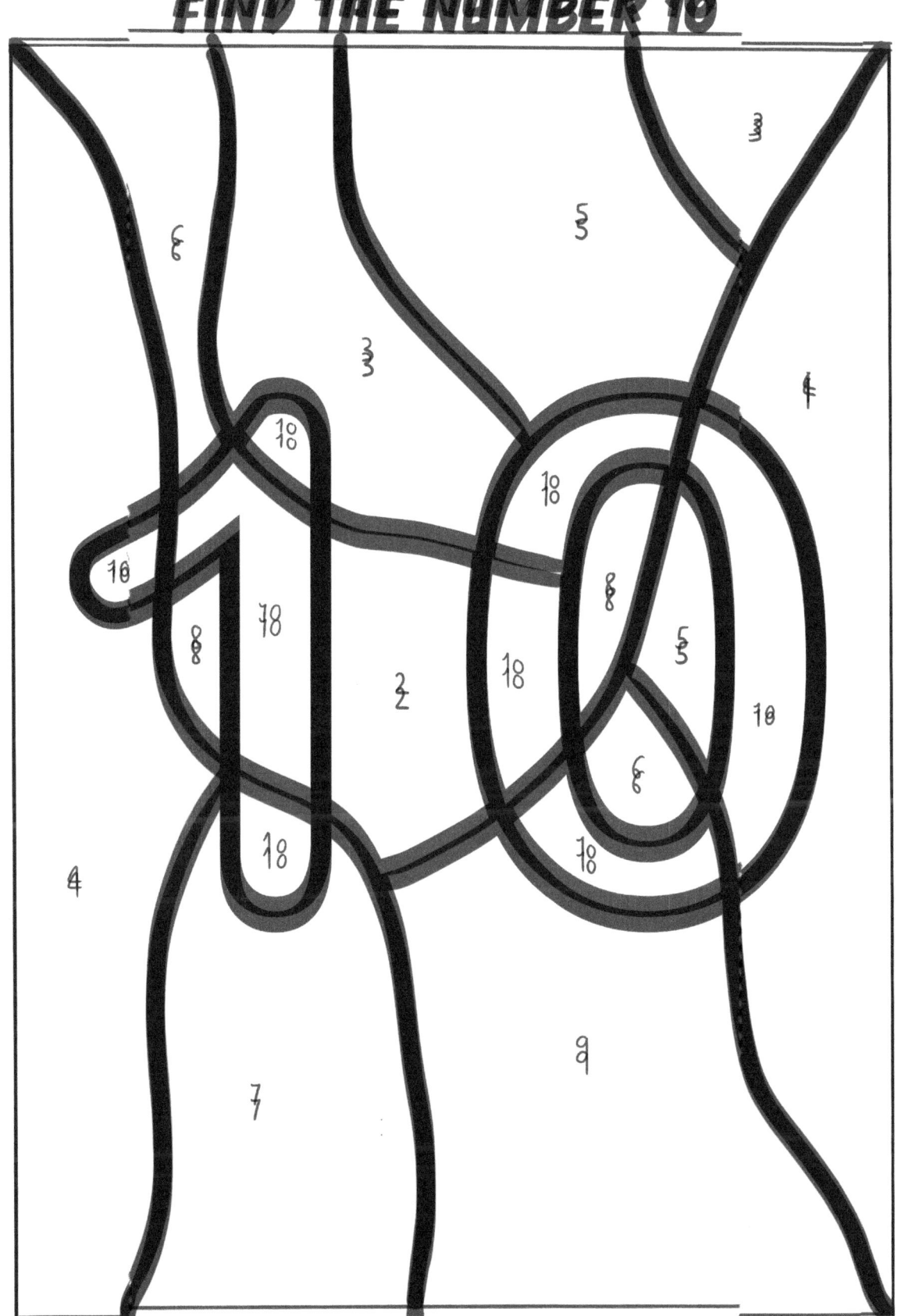